TEXT
MURRAY WALDREN
ROBERT RENEW

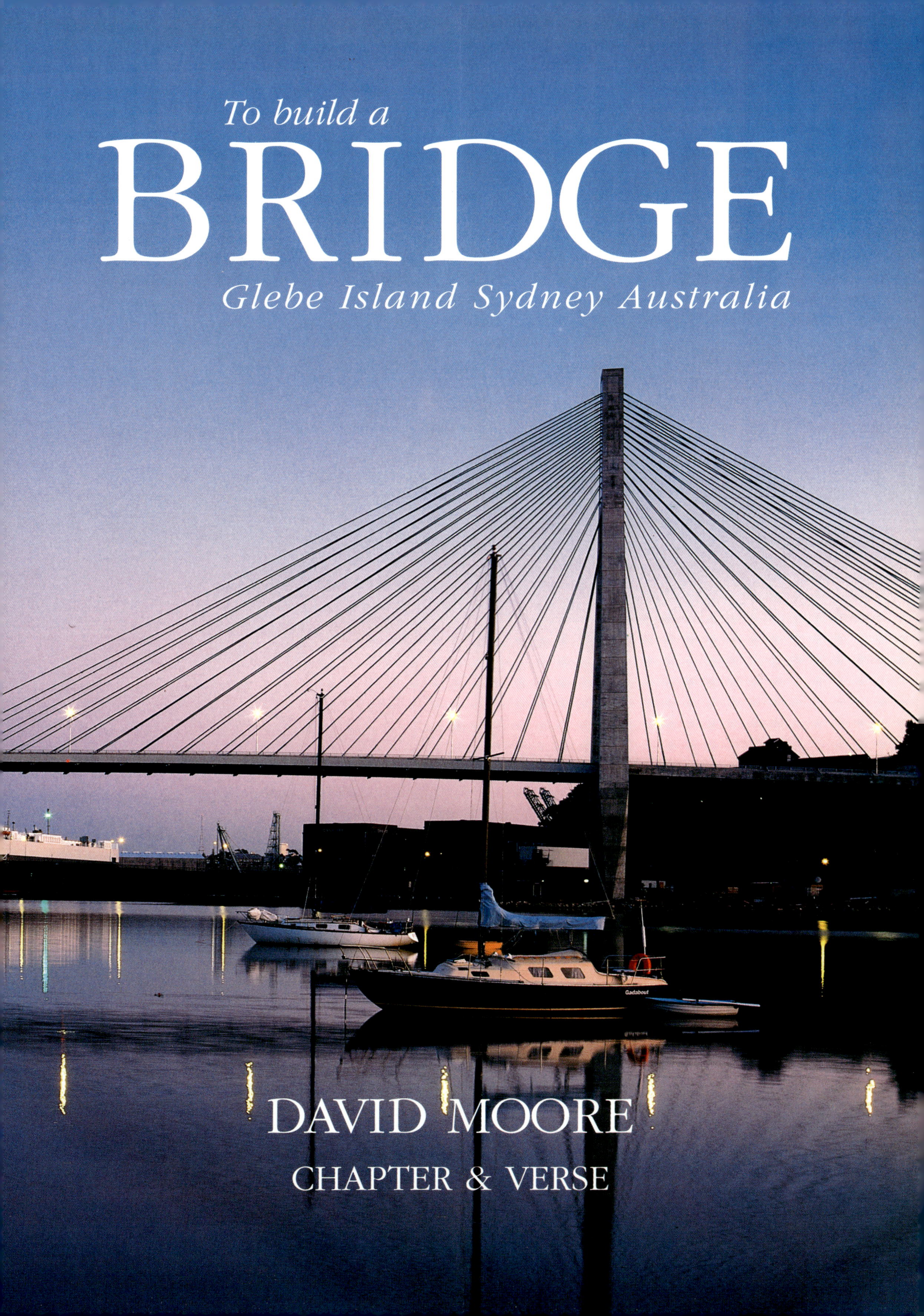

To build a

BRIDGE

Glebe Island Sydney Australia

DAVID MOORE

CHAPTER & VERSE

Half title: Survey graphic
Title pages: Glebe Island Bridge at dawn

Above: Carpenter's tool belt
Opposite page: Stay cable abstract

First published in 1996 by Chapter & Verse, an imprint of Wellington Lane Press Pty Ltd, 120 Wycombe Road, Neutral Bay, NSW 2089 and John Witzig & Company Pty Ltd, Mullumbimby, NSW 2482.

Designed by John Witzig
Edited by Carol Dettmann

National Library of Australia
Cataloguing-in-Publication data

Moore, David, 1927 – .
To build a bridge: Glebe Island, Sydney, Australia.

ISBN 0 947322 10 8.

1. Bridges, Cable-stayed – New South Wales – Glebe Island – Pictorial works. 2. Bridges, Cable-stayed – New South Wales – Glebe Island. 3. Glebe Island Bridge (Sydney, N.S.W.) – Pictorial works. 4. Glebe Island Bridge (Sydney, N.S.W.). I. Waldren, Murray. II. Renew, Robert, 1948– . III. Title.

624.28099441

Printed in Singapore by Toppan Printing Co (Australia) Pty Ltd

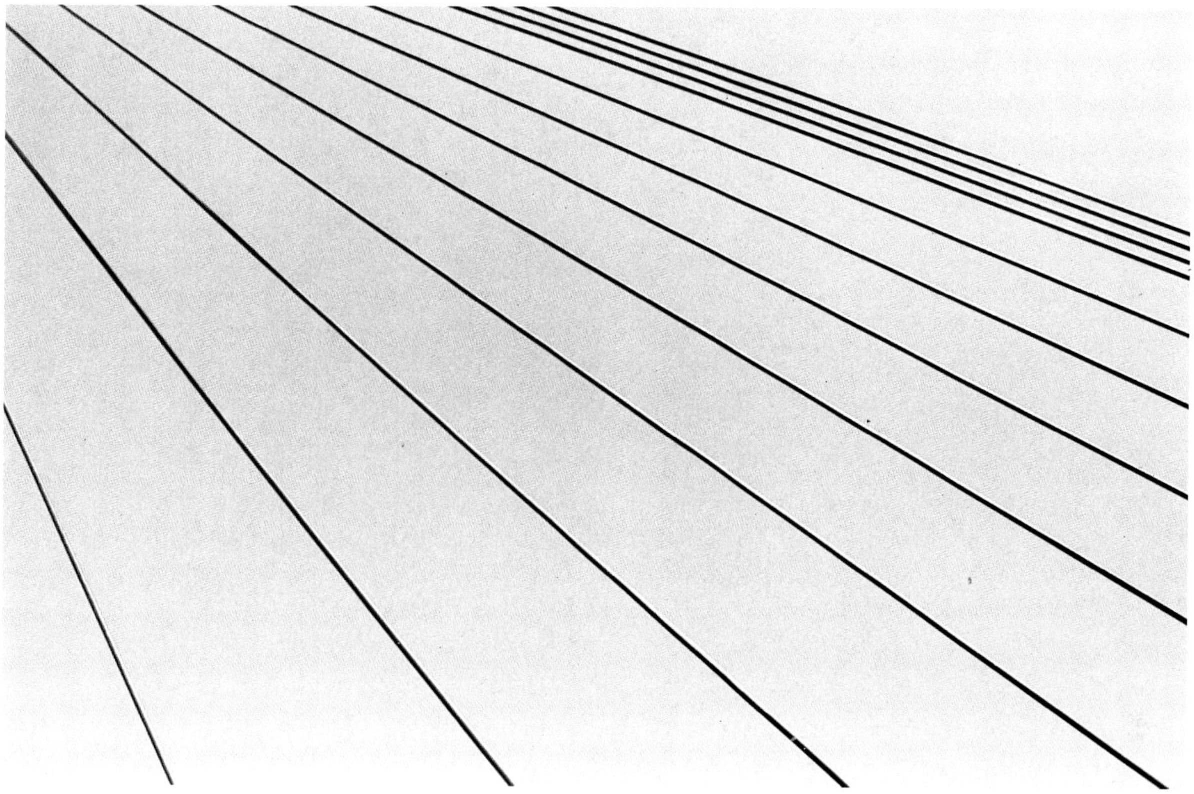

Acknowledgements

Grateful acknowledgement is made to those organisations and companies who have generously supported the publication of this book.

THE ROADS AND TRAFFIC AUTHORITY, NEW SOUTH WALES

BAULDERSTONE HORNIBROOK ENGINEERING PTY LTD
Major contractor

Austress Freyssinet Pty Limited
Supply and installation of stay cables, tower jumpform and prestressing systems

Maxwell Optical Industries Pty Limited
Australian distributors of Nikon camera equipment used in coverage

SMS Consulting Group Pty Limited
'Bridging the gap between business and technology through people'

The project was further assisted by a Fellowship from the Commonwealth Government through the Australia Council, its arts funding and advisory body

INTRODUCTION

TO BUILD A BRIDGE is a feat of daring.

Risk factors associated with erecting a precise steel and concrete span are ever present and there is a splendid bravado evident in the vault through space. Perhaps nowhere else in the field of engineering design can form and structure be so readily appreciated. In late twentieth-century bridges, an essential, functional reasoning allows no place for applied decoration or cosmetic overlay. Structural elements are left unclad, imparting an aesthetic purity which is a direct resolution of design problems coupled with a delicate balance of tensions.

Between 1933 and 1936 the San Francisco Oakland Bay Bridge was built at tragic cost as 23 workers lost their lives on the project. A 21-year-old photographer, Peter Stackpole, set himself the task of documenting the building of that bridge. He did it with consummate passion allied with an impressive ability to express the forms, graphic elements and workers' dedication. This body of work produced a supreme statement of photography.

When I saw Stackpole's pictures some years ago, my response was deep appreciation. The photographs were clear, sharp and vital, expressing the essence of the venture.

In 1975 I chanced upon the photographs Henri Mallard had made covering the building of the Sydney Harbour Bridge. The negatives were passed to me by Henri's son, Paul Mallard, to be printed for an exhibition and a book. Again I was impressed by the photographer's devotion to documenting an engineering wonder.

The photographs of both these bridges now have major historical significance. They speak of a time long gone and show, with perception and power, the work as it progressed. Neither photographer was content to make his coverage from ground level only.

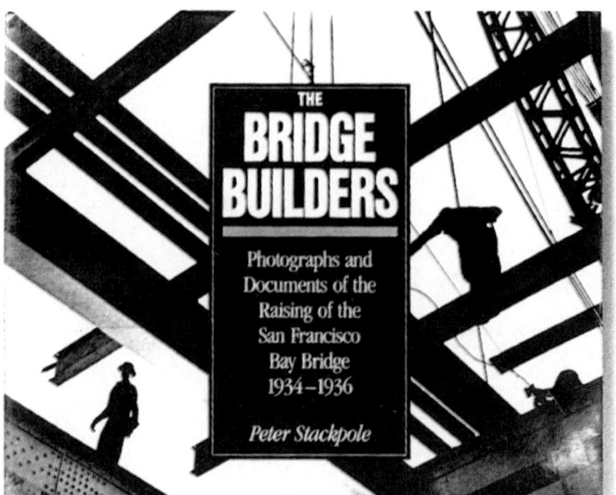

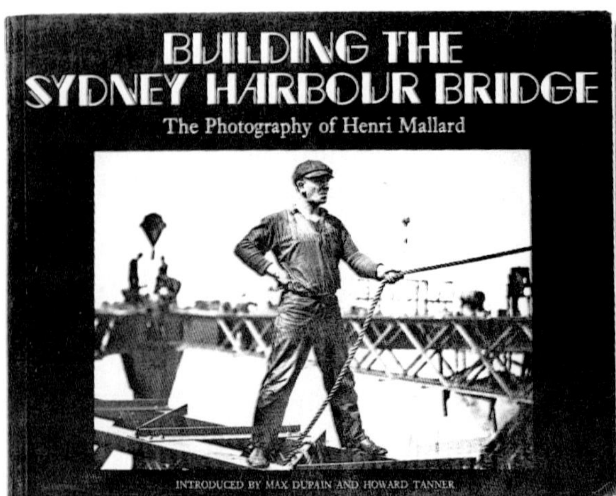

Each sought out the riggers, dogmen, steelfixers, riveters, signalmen and painters high above the water to make images that celebrated the perilous work.

As foundation work for the Glebe Island Bridge began I remembered the pictures of Stackpole and Mallard. The construction of a bridge of such magnitude only happens once in a photographer's lifetime and inevitably the photographs of the project will interest many people in years to come. With this in mind I decided to plan a three year period of work to concentrate on the venture.

The Roads and Traffic Authority and the major contractor, Baulderstone Hornibrook, were supportive of my project. Both agreed to allow me unrestricted (and unescorted) access to any part of the bridge during construction so that I could build my statement. It was important to be an independent observer – free of any restrictions imposed by commercial considerations or client brief.

When first visiting the site I felt like a new boy at school. I had met a few of the engineers briefly but the general work force were totally unknown to me. In time the men accepted my presence as part of the operation and showed me every courtesy. Not wanting to interrupt their work by asking for any set-up pictures, I photographed discreetly as they went about their tasks. It was not long before I understood the pride and dedication they felt in their work: much of the coverage seeks to express this attitude.

To see the bridge grow from ground level to a soaring construction of concrete and cables against the backdrop of the city and the Harbour Bridge was a joy. I looked forward to rising before dawn to watch the early light wash the scene and sometimes hurried to Glebe Point or Rozelle when impressive storms swept through from the south-west.

Occasionally when standing on the roadway deck, it was possible to feel the life force of the structure. Surveyors told me that the apex of the towers moved as the sun angle changed and movements could be detected underfoot as concrete trucks rolled up to deposit their loads. After the two spans were connected and the road surface laid, the many support cables were adjusted and tensioned much as a musical instrument is tuned.

Since mid-1992 until the official opening of the roadway in December 1995, many thousands of frames have passed through my cameras. During the final year the project possessed me totally for I was conscious of the fact that to miss an important point of construction would leave a vacant hole in the fabric of the coverage. If my photographs assist in celebrating the expertise, dedication and responsibility of all involved, my time will have been well spent.

David Moore
McMahons Point 1996

Western approach piers under construction, August 1992 ▶

 Reinforcing steel and formwork on western approach

Clockwise from top left: Worker with plank and crane tracks; carpenters prefabricating formwork; a load of reinforcing steel arriving on site

Steelfixer inside base of western tower
Opposite page: Reinforcing steel assembly in western tower base

 Jumpform screw jacks on lower tower leg

Leading hand carpenter's tool belt

 Laurie Chow, RTA construction engineer

Bruce Wilson, project engineer, and Simon Lillyman, administration manager, in Baulderstone Hornibrook Engineering site office

Meeting in site office between RTA and Baulderstone Hornibrook engineers. Sandy Buchan, Baulderstone Hornibrook engineering project director and Peter Wellings, RTA resident engineer at head of table.

From top: Peter Wellings and Laurie Chow; Sandy Buchan

Merv Francis, Scott Wyatt and Mario Perkovic adjust chains for lift of precast concrete panel
Opposite page: Placing precast panel in western abutment retaining wall

 Western tower and Sydney Harbour Bridge from Glebe

Above & top: Work on western deck stay cable guide pipe

Above right: Tony Figueredo carrying scaffold pipe over reinforcing steel on western approach span

From top: Dani Te-Nah-U on section of Kingshore support scaffolding; crane driver Joe Stelzer at controls of 4100 Ringer Mantowoc crane

Placing hooks for slinging of Kingshore towers ready for erection

Erecting first section of Kingshore scaffolding under western tower
Opposite page: Western tower seen through safety screen

Jumpforms on western tower

Opposite page: Western tower seen from eastern shore

◄ Riggers connect Kingshore scaffolding sections under western tower

 Western tower with Johnstons and Blackwattle Bays and city backdrop. The 4100 Ringer Mantowoc crane has a maximum jib height of 140 metres

Ivan Carapina, construction worker

Above: Bob Short placing reinforcing steel on eastern approach span
Right: Pre-assembling reinforced cages for deck edge beams

 Clockwise from top left: Mark Taylor, steelfixer; Bill Tams, boilermaker; Alan Breslin, steelfixer

Kingshore scaffolding in place supporting pier table under western tower

 Placing concrete on western pier table

Clockwise from top left: Joe Malacia, concreter; John Torrisi, concreter; Darrin Watson, concrete boom operator

From top: David Hardie, RTA inspector, watching as foreman Trevor Weller directs work on western pier table; John Baker signals to stop concrete supply on western pier table
Opposite page: Rod Collier and Bruce Cross manhandle stressing trumpet into position

42K13

Ken Kaye-Smith guides cables into heavy lift jack
Opposite page: Heavy lift jacks in place on western approach span

 The towers rising against backdrop of the city with Rozelle Bay in foreground

Formtraveller is manoeuvred into position before lifting to western deck

 One of the first multi-strand stay cables in position

Bill Jenkins attaching mono-strand jack for tensioning of individual cable strand in anchorage on western tower head

Hermino Bessa draws cable strand through polyethylene sheath
Opposite page: Construction of western deck with formtraveller in place

You find the house, we'll come to the pa

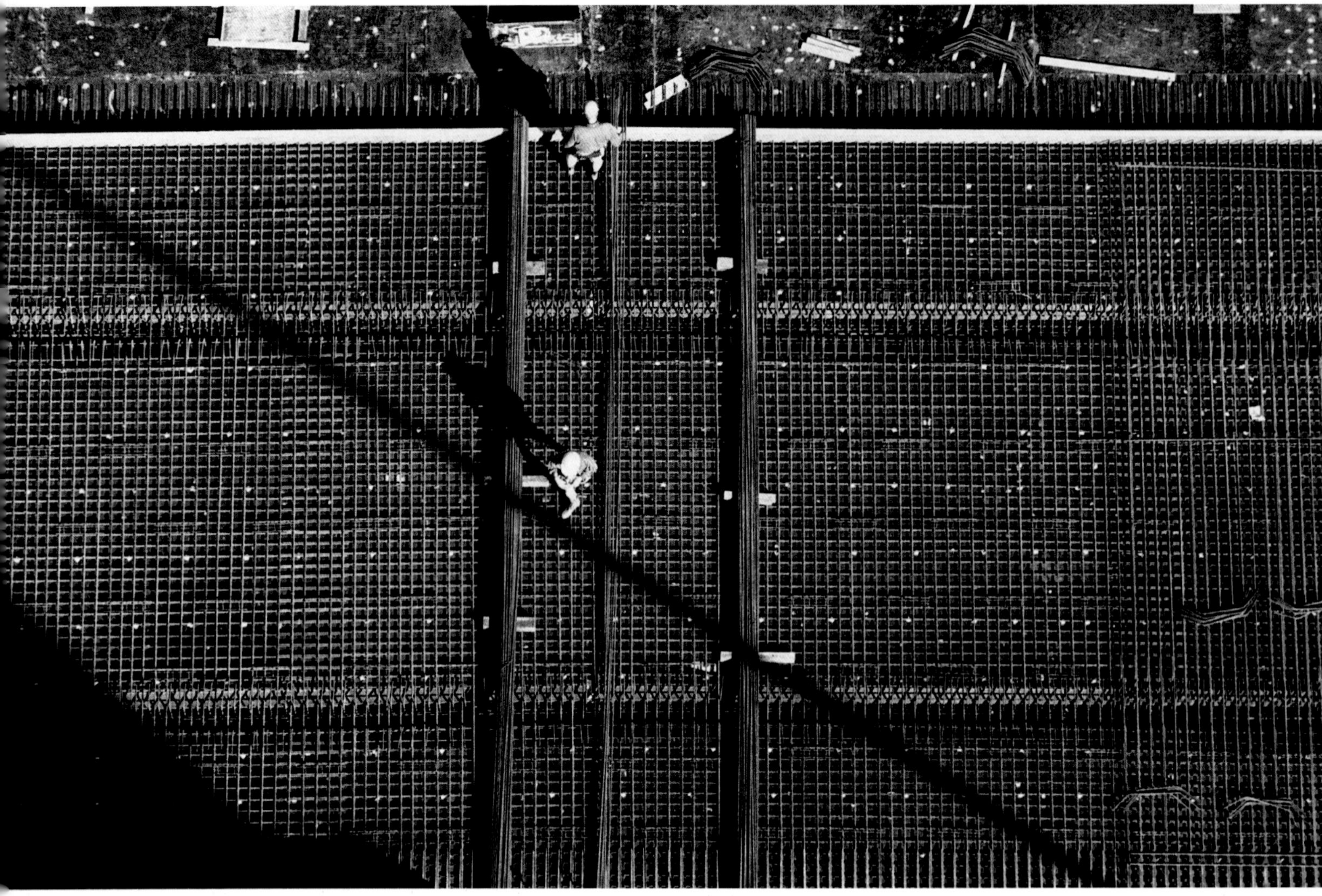

 Preparing reinforcing steel for concrete on western deck

From top: Concrete placing team on western deck;
Maafu Matavesi finishing concrete on eastern deck segment

 Kofeinu Mataresi signals to pump controller for position of concrete placement

Henry Malacia uses his feet to spread concrete on western deck

The segmental deck cantilevers out from the western tower supported by stay cables ▶

CANE-ITE

Eastern approach viaducts leading from Darling Harbour under construction

THE CABLE STAY SOLUTION

Murray Waldren

THEY HAD WANTED A CALM EVENING. Sunlight made the daytime inappropriate – it hit the spans at different times, causing them to expand and contract at uneven rates. A storm front with its sudden temperature drop would have been disastrous. Only a uniform temperature would allow a compatible, effective joining. If it had been designed to order, that July 24 night in 1995 could not have fitted better: unusually crisp, it had more than a hint of freeze in the wind.

On the bridge decks, work lights blazed with day-bright intensity. There purely as function, not form, they lent a theatrical surrealism to the event, an industrial stage-set of men and machinery some 30 metres above the shadowy swirl of Sydney's Johnstons Bay, at its intersection with Blackwattle and Rozelle Bays. The decks themselves, looming from darkness beneath sculpturally silhouetted towers at either end, seemed to arch yearningly – fingers straining to touch in the spotlight. Yet for all the high-tech sheen, there was something primordial about this middle-of-the-night atmosphere: the queue of pachydermic concrete trucks awaiting connection to an elephantine trunk of the high-hosed distributor, the Stonehenge of formwork and jutting steel rods, the arcane rituals of workers in firelight.

There was symbolism of a different kind in the shush of traffic passing the nearby Olympic-muralled silos and sweeping over the low-slung Glebe Island swing-bridge. Along with its sisterly Pyrmont Bridge, it had been among the first electrically-operated opening bridges in the world, a marvel of its day when it was inaugurated in 1903. Even then, it had not been the first bridge in the area, progress dictating it should supersede another built in the 1850s to provide access to the then city abattoirs at Glebe Island. In the 92 years since, the world had changed beyond its designers' wildest expectations and for some years it had been well past its use-by date. Effectively at least, if not emotionally. Regular users and locals still held it in fond regard, and if most had at some time cursed its erratic opening for delaying their daily rush, it was still a touchstone with a less timeworn past. At peak times, however, its shortcomings condemned it in a city asphyxiating with vehicles.

Plus ça change... As the new bridge soared above and beyond the old, seagulls confused by the untimely lights swirled like a disturbed swarm of gnats, adding an eerie resonance to the night. And away in the middle distance, city neon and streetlights shimmered in a half-haze, their patterns on the water Turneresque reminders of just how beautiful the city's swirl of bays and coves could be.

Suddenly it was all hands on decks. Site engineer Bruce Wilson, general foreman Trevor Weller, superintendent Jeff Young and inspector David Hardie, all veterans of the project from Day 1, gave let's-go signals to the 10-man concrete team. Its united nations membership reminiscent of the legendary Snowy River Scheme workforce, the crew was phlegmatically nonchalant in the manner of all working gangs as they inspected the final trench of layered reinforced steel grids. A discreet distance away, huddles of executives, invited VIPs, PR representatives and print and TV journalists hunched self-consciously in borrowed yellow raincoats, steel safety boots and hard hats. And if some among them wondered aloud just how exciting watching concrete being placed could be on a freezing night in July, they also suspected they were about to witness urban history – a small but significant phase in creating the controversial Glebe Island Bridge.

Nearly ten years after the first design lines were drawn at the NSW Roads and Traffic Authority's headquarters, some six years after Leighton began the work of hammering 112 pile casings up to 35 metres into the solid sandstone base to support the towers, and after considerable turmoil and doubt, Australia's largest span cable-stayed bridge was becoming reality with the joining of its eastern and western spans. The night's placement may have been tiny – just 26 cubic metres of concrete – but it was adding weight exponentially greater than its 65 tonnes to the whole.

As concrete pumped into the formwork, leading hand Astrubel Eusebio and his team swung into their well-practised routine. Already they had placed 52 sections above the bay, some 8580 cubic metres of concrete on the cable-stayed deck alone; this night's work was relatively simple. To the casual observer, the activity verged on organised chaos; to the crew and the deck staff, it was just another day at the office. Well, perhaps not quite. For some, there were unexpected end-of-an-era echoes. "It's a funny feeling, actually", Trevor Weller admitted as he watched the formworked channel slowly fill. "There's a strange sense of loss about this. It's almost an anticlimax after all the years of working so hard and so single-mindedly towards this."

The job settled into its own rhythm, the gaggle of observers into degrees of distraction or expectation. Except, that is, for the discreet figure of David Moore. Beneath his personalised hard hat, his bulldog features were all concentration as he darted in close to the bustle, probing for an edge, the right angle. His movements were lithe, emphatic. And although it was impossible to see art, or even artifice, in what he was shooting – a work site melee of concrete, cables and mild confusion – his photographs proved otherwise. This was experience and intuition working together; the visitors were seeing a concrete pour: he was seeing opportunity, a mother-lode of photographic possibilities. "I don't know that I'm ever very analytical or conscious of the way I'm shooting pictures", he told me later. "A lot of it has to come from the subconscious, through a mixture of understanding and ability. If you make it too conscious, it becomes laboured – you just have to feel it, and respond automatically."

Perhaps more telling of the man and his approach was the respect the workers showed him amidst the drama of the night. To me, they talked with quiet eloquence of his being a good bloke, someone who understood, who shared their pride, their effort, even their sense of destiny at working here; to him they showed the acceptance of unselfconscious disregard, professional to professional.

"I THINK THIS WILL BE THE MOST IMPORTANT WORK I'VE UNDERTAKEN", Moore told me early in 1993 of his then just-started project. At the time I was interviewing him in the lead up to his 50-year Sydney Harbour retrospective at the State Library of NSW, and I was surprised, even a little sceptical, that against such a body of achievement he should rate a bridge construction shoot so highly. I shouldn't have been – but to understand the weight of his statement, you need to remember just how significant a photographer he is.

Simply put, his photographs have, over five decades, helped define Australia's visual vocabulary, and through that, our referential history. His many commissions, books, exhibitions and even individual photographs – such as the famous *Sydney Harbour Bridge* (1947) or *Redfern Interior* (1949) or *Migrants* (1966) – have pinned down defining moments of our common experience, fixed perceptions of place and self in our minds. Through his international exposure his work has been a valid and valuable aspect of our cultural emancipation – showing our reality, on our terms, overseas.

The project needed someone like him to get involved, someone clear enough of other commitments to do it. It's one thing to have a contracted journeyman photographer shoot time-lapse records, another to have someone concentrate intently on a project for more than three years, to look at it from an individual view as a dramatic situation to document for the future. Equally important was the need for that photographer to be free of commercial considerations and client briefs.

This work could also be seen as a culmination of everything he had been trained to do over the years. It had special demands, requiring the insights and skill gleaned from decades of architectural photography, the ability to shoot heavy metal industrial commissions on-site in all conditions, to capture the fleeting as well as the timeless. His photojournalism was internationally renowned, his landscapes part of the local idiom. "I don't know of anyone else with that sort of experience who was also in a position to concentrate on it as I was", Moore agrees. "Knowing it was a three-year plus project, you had to be able to make a commitment. And it was a big commitment, both financially and in time. Halfway through, I was lucky enough to get a Keating fellowship. That helped alleviate the financial drain but it didn't change my input – I didn't shoot any more than I would have normally. Still, I have to say it has been the longest single project I've ever worked on with such intensity."

Visually, he found the experience almost constantly inspiring. "I love graphics in photography, and all this skeletal steel and angularity was right down my alley. It's been like watching a huge piece of sculpture develop, which essentially is what a construction like this is. Yet here I had the opportunity not just of seeing the finished work in an exhibition but of watching it pull together. From a photographer's viewpoint, it could not have been built in a better place, with the city backdrop, the sweep of bays, the wide overhang of sky, the Harbour Bridge in the distance, even the concrete silos across the road..."

PHOTOGRAPHIC CONSIDERATIONS ARE ONE THING, politics another. Few public projects of recent times have experienced the uproar that swirled around this bridge's construction. True, it did not have tens of thousands of protesters blockading it, as with the Sydney Airport Third Runway development, but it did attract a persistent chorus of dissent throughout. There was actually little public response to the design or environmental impact statement until March 1992 when the towers began to rise. Then opinion on its desirability, necessity, aesthetics and size became largely antagonistic. Architects decried its design, commuters and motoring organisations questioned its then projected six-lane width. All this, they asked, for just two traffic lanes more?

For others, it was further evidence that Sydney was "ruled by the car", that the interests of unimpeded traffic flow outweighed all other considerations. Many felt the bridge's role in linking up freeway continuity was less a vital community service than the final step in a Los Angelisation of the city. Still others complained that its construction would promote the "urban doughnut", encouraging through improved access and reduced travel time a residential movement away from the CBD and inner-city suburbs to suburban commuter centres.

Community and environmental opposition in its inner-west home ground was especially loud, based largely on an alleged "boys-with-toys" approach that smacked, they said, of overt monumentalism. Some opposed the bridge's "inappropriate" siting in the context of the bays, its dominance of natural aesthetics, its probable ruining of views, the noise factor...

There were complaints that the design was unnecessarily costly, that when finished the bridge would not solve the traffic bottleneck but merely push it further along the road. Midway through its construction, more controversy erupted when the only two cement ships requiring access under it ceased operating in the bay. Their contracted long-term access rights had been a factor on the bridge's final design, and its cost (a State Development Department report had found it would be significantly cheaper to raise the bridge – and absorb the extra cost incurred in that – than to retain a low-profile design and pay an estimated $50 million compensation to relocate the companies involved). Conspiracists alleged the decks had been designed with such high clearance to facilitate future construction of a high revenue-earning 1000-berth marina in the area. Such a proposal had been mooted before, only to fade away after opposition from potentially affected locals.

Newspapers and local politicians loved the debate; it was good copy, good leverage for political point-scoring, good fodder for dinner table discussion. Emotion at times ran high, and hard. One of the more persistent critics was leading architect Philip Cox, who wrote in *The Sydney Morning Herald* in December 1993 that the bridge "does much to destroy the scale of intimate harbour inlets and Pyrmont Bridge and White Bay". He was not alone in his condemnation – other critics ranged from the Prime Minister, Paul Keating, to Sydney's Lord Mayor Frank Sartor, to the State's Minister for Transport, Brian Langton (he described it once as a monument to over-engineering), to Larry Hand, the former mayor of the Leichhardt municipality that includes the bridge-affected suburbs of Rozelle and Balmain.

The issues raised were, and still are, germane to large-scale city construction in the late twentieth century, to how life in and around the city is lived and where the emphasis on public and private access to the CBD falls. In many ways, the bridge – so omnipresent in its design and siting – became a cathartic transference for the frustrations and apparent powerlessness of people who saw the city becoming more alien and less communal, who believed that city planners understood the macro but were micro-illiterate. As with all such situations, there were half-truths and exaggerations, and emotion coloured perceptions. But the points of contention remain valid, and the fact of this bridge's existence does not diminish their need for resolution in future public works.

Not all the dilemmas can be sourced to the bridge itself, particularly if its construction is viewed as a logical extension of a freeway policy introduced – no matter its appropriateness – decades before and which was choking itself to ineffectiveness at a major gateway to the city. As well, some of the opposition was aimed more at the inelegant "spaghetti-way" of ramps leading to the bridge than at the bridge itself (although the two were inexorably tied). This was small comfort to the RTA – accusations of a bovver-boot sensitivity stung an organisation apparently keen to make-over its public perception after inheriting a disastrous image from its earlier bureaucratic incarnation as the Department of Main Roads – or to the bridge's constructors, Baulderstone Hornibrook. What they both saw as a huge opportunity, as a technologically exciting, cutting-edge development that would give the city an exciting landmark while improving its lifestyle, was being publicly repudiated.

"For several years there was a feeling of being in the bunker", agrees the RTA's Chief Bridge Engineer Ray Wedgwood, "when every comment about the bridge seemed negative. We used to tell the troops not to take the comments personally but it was a bit ... disturbing, and personalised. The main thrust was that RTA engineers were on an ego trip or gold-plating things, that there were cheaper solutions. And I must admit that to look at the site in a cursory way, you could be justified in saying there was a cheaper solution. But when you look deeper and take in all the immovables, it's very obvious that ours was the logical – and the best – solution."

Public opinion began to show signs of swinging around once the structure became more than two phallic towers and gained some decking. There was a more noticeable surge in acceptance after the cables were strung. These had an immediate effect in lightening the mass, elevating the bridge from the earthbound to the more ethereal. Today, after the bread-and-circuses excitement of its completion and opening, acceptance is considerably warmer and more widespread. It's still not unanimous – some architects, engineers and artists continue to argue alternative solutions; councils and motoring bodies praise the Bridge but point to difficulties further along the track – yet the public generally seems pleased. The protesters are philosophic: it could, one leading figure told me, have been much worse. Translated, that's almost a vote of confidence. Few would suggest the Glebe Island Bridge will ever match the Sydney Harbour Bridge in its symbolic or iconic power, but it is now seen – along with the Gladesville Bridge – as one leg of Sydney's Towering Triumvirate of big bridges, an essential element in the city's Olympic Highway for the 2000 Games.

That alone gives the constructors satisfaction. As Wedgwood says, there has been more than the usual excitement within the RTA over this structure: "That's to do with the accomplishment of it, the aesthetics and the technology. And there is, I admit, a feeling of relief that we're now able to put our heads up and feel vindicated."

Ray Brown, the RTA's Project Director, is more forthright. The Glebe Island Bridge was "his baby" and he felt keenly that the sniping and criticism of it during the construction was "unjustified, and unfairly detracted from the skill and dedication of the workers and what they were accomplishing". The huge community response at the bridge's official opening assures him, he says, "that the commitment, enthusiasm and effort of all those involved has led to a very functional and largely popular project".

THE GENESIS OF LANDMARK CONSTRUCTIONS like the Glebe Island Bridge is never clear cut. Their evolution quickly becomes obscured in a misty territory of cause and effect, easily as labyrinthine as the skeined byways the bridge has been designed to replace. Unsurprisingly, pragmatism rules. More endearingly perhaps, pragmatism allows occasional unanticipated flights of fancy.

Simply put, the Glebe Island Bridge exists because Sydneysiders are car-obsessed. It's an obsession that took hold almost from the automobile's creation, has remained unaltered since and in the process altered the very city and the way people live there. More cars meant more arterial roadways. More roadways meant increasingly isolated islands of population within a gridway heading to gridlock. Hence the philosophy behind freeway development. Cynics might question whether, in a traffic sense, the free of freeways translates to the free of freedom, but that's certainly been the drive behind the coordinated highway construction. Perhaps belatedly, past city elders saw the future, and the future was an approaching meltdown. The solution, they believed, lay in a more-is-less efficacy: more effective major routes, less imperative for diversion through suburban "shortcuts".

It could be argued that the city became locked into a major highway system, and all that implies, once the Harbour Bridge was opened. It's all but forgotten today, given the universal love affair with this meccano-and-stone meld, that the Harbour Bridge itself generated controversy at the time – its history was a long one of proposal and rejection, and when it was finally built, prominent critics averred that it "had surely ruined the Harbour for ever". The city, they harumphed, would never be the same. They were right, of course, but not in the way they meant.

The quickly iconised Harbour Bridge provided speedy, direct access, and as a by-product promoted rapid colonisation of then relatively isolated "villages" to the north and north-west. The city, swallowing suburb after suburb in its outward expansion north, south and west, became irrevocably committed to an expressway philosophy when the Cahill Expressway cut across Circular Quay; the metamorphosis into a freeway-feeder precinct became inevitable once traffic and population exploded, and the need for rapid deployment of people and resources became tantamount.

Whether or not expressways are the most effective traffic solution, or whether traffic expands to clog such arteries, are different social arguments. But once the Darling Harbour freeway distribution was proposed, it made no sense to urban planners not to extend westwards to feed traffic in and out of the city with maximum efficiency at its last point of serious congestion.

Actually, early plans for the western freeway link did not include a bridge at the Johnstons/Blackwattle/Rozelle Bays area at all. Initial proposals had it bypassing the waterway altogether, cutting more or less directly from the city through inner suburbs and then out. It's ironic, given the environmental concerns over its construction, that the Glebe Island Bridge came about largely because of successful environmental agitation and threatened green bans against the original suggestion.

Tom Uren, former Labor federal minister and Balmain boy (his father worked as a builder's labourer on the wheat silos at White Bay), claims, only partly tongue-in-cheek, that he is responsible for the Glebe Island Bridge's construction. As the Minister for Urban and Regional Development in 1974 he and his department prevented the western freeway link from extending over Wentworth Park and on through Glebe, Annandale and other established suburbs. The transport division of his department carried out a study of alternative routes. Two consultant reports were undertaken, the second completed in 1976 after the Whitlam government had been removed from office. At Uren's urging, however, the newly-elected State premier Neville Wran appointed a committee to investigate inner-city freeway strategies. That committee ultimately endorsed the first consultant proposal, which included a bridge at Johnstons Bay. "The new freeway that crosses Pyrmont, Glebe Island and White Bay has", Uren writes in his autobiography, "changed the whole inner western Sydney freeway system. We can say with some pride that our actions saved suburbs ... from being dehumanised and scarred with a wide swathe that would have cut them in two."

Cut to 1984. "I was leading the design team that first drew lines on paper", Ray Wedgwood, a 32-year veteran of the RTA, says. "We started off when the need for the bridge was finally established with a Moonie Moonie Bridge-style design with the legs cut off. I was working with Ken Wheeler and others in a project team for Ray Brown and at the time we thought we could sink the piers into the harbour and build islands around them. When the Electricity Commission heard about it, they told us they had a high-voltage line there which supplied about two-thirds of Sydney. Upon investigation, we found this cable was in a big arc upstream."

The RTA explored other forms of construction around piers to give protection to both bridge and cable, but the alternatives were unacceptably expensive. Negotiations to relocate the electricity cable pointed to a cost in the multi-millions. Then came the discovery that there were also major telephone cable networks in the bay. "The potential costs were so high by then", says Wedgwood, "that it was suggesting to us that we should get out of the water, always a good thing to do if you've got ships around. We looked at the possibilities of the full span because we were getting beyond what was thought to be economical for balanced cantilever bridges." In the end, he says, the cable stay became the obvious and only solution.

"That, however, became a source of some controversy", agrees the bridge's Resident Engineer, Peter Wellings. "The original proposed bridge had a significant visual impact on the city, its replacement an entirely different one – I'm not saying an adverse impact but it is substantial. I actually came to the project when it became apparent the piers had to come out of the water, and the job then became a massive undertaking, just in coordinating the environmental impact statements, the design, the funding, the necessary land acquisitions, the drawing up of contract documents, the letting of tenders ..."

One defining element was the clearance above the water, which was determined after detailed negotiations with the Maritime Services Board. Originally, the MSB had decreed that the clearance should be 39 metres (as some perspective, the Harbour Bridge has a clearance of 53 metres). The RTA, correctly enough, believed the public would reject such a dominating elevation; it also questioned the necessity of such a height. Eventually, the MSB agreed to a 27 metre clearance as being the minimum it could accept, given the needs of vessels around the area. With the height of the decks and the span of the crossing set, the physical dimensions of the bridge became "a matter of geometry", says Wellings. "Once the defining elements were established, we had to design according to structural requirements. There was no excess in anything we did – it was purely an engineering problem."

"When the parameters were defined, it became necessary to make it aesthetically unified", says Wedgwood. "RTA principle architect Neville Fielding and engineer Brian Fredericks did that work, and determined the shape of the towers. Their brief also included analysing the proportions – admittedly more from a structural efficiency point of view than aesthetics. Many parameters of this were fixed, but there were still many variables. We also investigated various combinations of materials for the structure; we canvassed widely, including in the US, for advice and information."

If it were being designed today, he says, it "probably wouldn't be so high above the water – the absolute theoretical minimum height the MSB would permit is just under 20 metres to allow a ferry to come through, which would have made the towers about 7 metres lower. But the fact is we had to design it for the conditions that existed then, according to parameters that could not be altered. And while the shape of the towers did allow options, and potential savings in say, a goalpost shape, my belief is that it was well worth paying a small premium to get the lozenge shape we have, particularly on a structure so close to the city as a backdoor to the CBD and as such an integral part of the skyline."

Project Director Ray Brown, however, insists that the clearance of 27 metres is still essential to provide access for all recreational vessels. "Even *Gretel*, the famous America's Cup challenger which uses the bay, would not be able to pass underneath with any reduction in clearance."

Wedgwood always "had a very strong belief in the appropriateness of the bridge itself. There was a lot of pressure on me at times to say there was a better way, [but] I couldn't because I didn't believe there was. There was nothing artificial in the way we arrived at the solution. Sometimes you can be involved in design and follow a project through and still have disappointments. I have no such regrets with this – the plans came together and the builders, Baulderstone Hornibrook, enhanced our design by using construction methods we hadn't envisaged. They've built it much smarter than we imagined – the way they've erected the towers, put together the anchorage for the stays, are examples of really clever engineering."

The efficiency of the workforce, and the innovative techniques of Baulderstone Hornibrook in the construction, actually spoiled a statistical symmetry. Sydney Harbour Bridge, the first of the city's trio of landmark bridges, was opened in 1932. Gladesville Bridge was officially opened 32 years later and Glebe Island Bridge was itself to open in late 1996. Such coincidences, though, are more beloved of poets and numerologists than engineers; that the bridge was finished more than a year ahead of schedule and under budget was poetry of a different kind to its builders. It sounds like the company line but there's no doubting Wellings' personal pride when he says that "We had to work very hard to meet all the needs of the stakeholders in the project, and that we succeeded is due absolutely to the teamwork and commitment of the workforce. They were dedicated to bringing it to reality, and they did. Above all else, we're most proud that there wasn't a single serious injury on the site during the construction (the Harbour Bridge, for instance, claimed 16 lives in its building) nor one day lost because of industrial dispute. "No one person can claim credit for the project – there's been a huge team that has lived through it – but in common we all saw this as a chance of a lifetime, an opportunity to fully extend our skills. And it was."

FROM THE AIR, SYDNEY LOOKS DECEPTIVELY FRAGILE, its bays and estuaries delineating a tide-torn lacework of peninsulas and promontories. All around, the sea dominates, threatening as it crashes against cliff or swamps sandy inlet or surges within harbours to wash away the land's tenuous claims. Even as the city's mass solidifies into a phalanx of house and highway, it is generously veined by rivers, creeks and watercourses. Only from the air can you appreciate just how much it is a city of waterways. And how many bridges it has linking cape to point, community to community. They have been built for traffic, for trains, for water and gas and power, for cyclists, for pedestrians. From the first Tank Stream Bridge of early colonisation, a plethora of bridges has been built, from the simplest "split-log roadway" to the most-involved engineered structure. In design – or by lack of it – they have little commonality other than function. Instead they constitute a living museum, a tangible history of the sensibilities of the times in which they were built. How else can you equate the castellated fantasy of stone that is Northbridge with the Bauhaus-like pragmatism of the Spit Bridge? Or the curvaceous arches of the Gladesville Bridge with the airy linearity of Parsley Bay's pedestrian walkway? Or even the self-effacing sweep of the Roseville Bridge with the angular symmetry of the new Glebe Island? By their idiosyncrasies and individualities, they help to define both the city and the harbour from which it radiates, adding their own pictorial theatrics to an inherently dramatic landscape.

Which is where David Moore comes in. As a boy, he became infatuated with Sydney Harbour's subtle moods, singular light and ever-changing bustle. His return home in 1958, following a successful seven-year stint in the photojournalistic proving ground of Europe, "rekindled that love affair" which has endured now for close on four decades. Throughout, he has recorded and interpreted the city and its harbour's many transitions with singular dedication. It's no surprise then that this decade's most significant "face-lift" to the waterside should have snared his attention. "I'm fascinated by change in cities", he admits, "and connected to it by the work I did in the 1940s. Then, the shots were simply exercises to develop my photography, but marvellously today those images express a time no longer there. The realisation of the importance of those images is partly what brought me to Glebe Island Bridge ... 50 years from now, the same sort of values will apply."

The images in this book are much more than a serendipitous coming together of dramatic site and natural sights. They are a melding of emotion and interpretation, of analysis and insight. "It became obvious very early on", Moore says, "that I shouldn't try to force things. The work became more a responding to what the bridge is. If I let it dictate terms, it rewarded me with its own geometries, physicalities, angles and arcs."

Out of this mass of steel and concrete, he has wrought abstracts of astonishing power and quixotic simplicity. He has juxtaposed the structure and the landscape with painterly perceptivity, suggesting at once both the enduring life of nature and the temporal nature of life. But it is his shots of the workers, whether antlike against a towering superstructure, en masse or individually, that add an extra dimension to his vision. The images invest the ordinary with an extraordinary insight into the dignity of labour, without trivialising or romanticising. There's a rapport with work and with workers here which raises the pictures beyond the record into the emblematic.

"After a while", Moore says, "instead of pushing myself to go down there, I found the bridge drawing me. It became like a hit; if I didn't get it frequently, I felt I was missing something. It took about 18 months, I guess, to get to that point ... and that was largely due to my relationship with the guys working on the deck and towers. I'm not a naturally gregarious person and despite my extensive industrial experience, I felt a bit of an invader, an alien on site. Once I realised I was accepted as part of the team, things began to feel really strong, rich and right. From there, the impetus and vitality started to become the engine of the project – it developed its own momentum.

"And I particularly wanted to convey the dedication of the workers in their job. On this construction, most had a real, heartfelt involvement. Although there are immense machinery and technology issues, its core still comes down to an intense manual labour element. I like being on a location where workers are doing a tough job – I've always found it produces valid photography."

No less valid, he feels, is the aesthetics of the bridge. "I can't accept the criticism some have made of it. There's so much to like about it, about its elegance – the tension generated by its springing off a narrow base, the double transparent pyramids you get in elevation ... The more contact I've had with the bridge and with the engineers on the job, the more I respect what the designers have done. Initially, I thought I had to keep an open mind until I finished as to whether the design was any good or not. But halfway through I felt I could confidently say, 'yes I like it, I like so many things about it'. Now I'm convinced it will become a city landmark in its own right. And future generations will look back on the arguments around it as we look back on those who decried the Opera House when it was being built."

Murray Waldren is a Sydney writer and Deputy Editor of The Australian Magazine.

Survey mark on deck

 Favco support towers in place during construction of western tower

 Concrete from truck is pumped to deck level through boom

Riggers working on Kingshore scaffolding

 From top: Stay cable anchorage assemblies; stay cable deviators

Western tower from east

FIRE EXTINGUISHER

Placing concrete in a segment at end of western span
Opposite page: Rigger on Kingshore scaffold

Construction graphics and details
Opposite page: Eastern tower at sunrise

Lampson
AUSTRALIA
PTY. LTD.
CAUTION
SERIAL No 48289
PLANT No 435
Lampson

 Prestressing ducting ready for placement

From the old Glebe Island Bridge on a winter morning. Western deck nears centre position

Above & opposite page: Steelfixers Tony Viegas, Reece Francis, Richard Harris and Michael Fisher, working on reinforcing cage around cable guide pipe

The western deck extends to central position of bridge 30 metres above the water
Opposite page: Peter Cole hoists a load to western tower head

From top: Eros Constandinos, boilermaker; Kingshore scaffolding ready to be moved by barge to eastern side of tower
Opposite page: Dawn light on the eastern tower with scaffolding about to be moved

 Above & opposite page: Lowering formtraveller onto barges at completion of western deck main span

Peter Woods working on jacks as formtraveller is raised
Opposite page: Formtraveller weighing 262 tonnes being raised by heavy lift jacks into position on eastern deck

Above & opposite page: Formtraveller rising to eastern deck

 A summer storm passes over the bridge and city

A worker climbs to deck level up access stairway near eastern tower

Opposite page: Eastern tower nearing completion

 Mark Taylor and Craig Scott on skidbeam with background of Kingshore scaffolding

Clockwise from top left: Tony Perin, leading hand carpenter, measuring distance between skidbeams; Joe Marton, rigger, moves cherry picker into position near eastern tower; Glen May, dogman, steadies a crane load before placement on barge; Craig Scott taking survey sighting

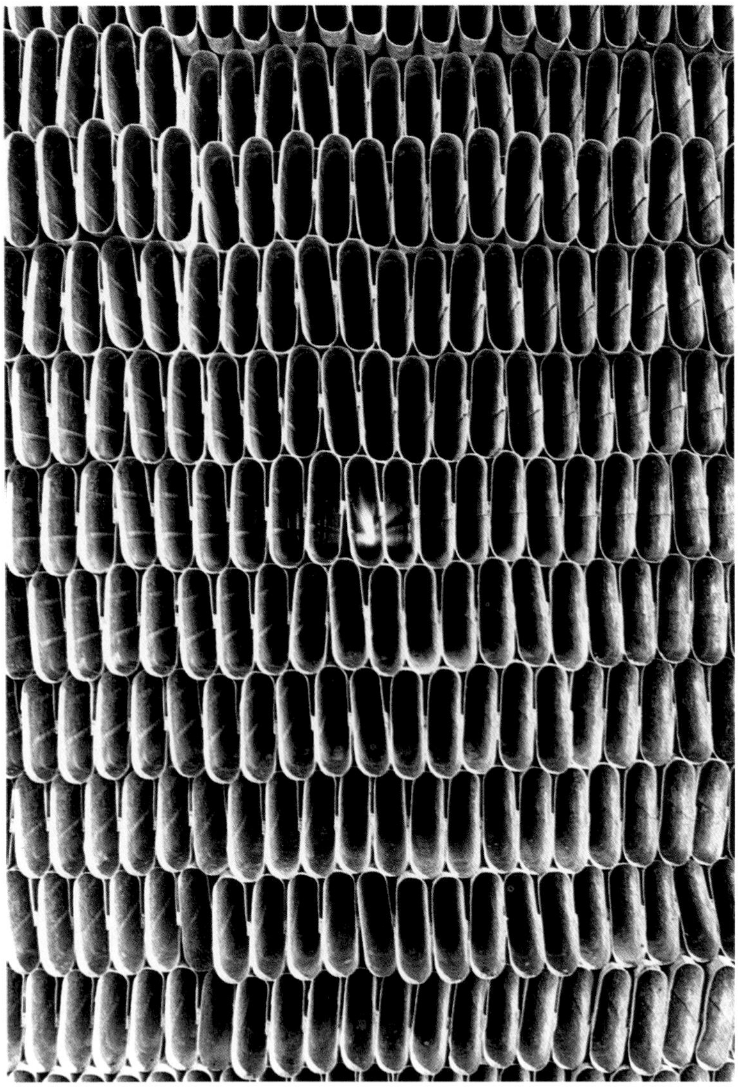

From left: Stack of ducting for post-tensioning of deck; safety screen fabric on stairway

Dave Latin positions a steel member before fixing

The bridge with deck gap narrowing seen at dusk looking south-west ▶

 Above & opposite page: Placing of curved steel girder on eastern approach viaduct at night

Heavy-duty cranes lower 42 metre long open box girder weighing 44 tonnes into position to rest on concrete headstocks on eastern approach viaduct ▶

SPINNING A WEB

Design and construction of the new Glebe Island Bridge

Robert Renew

STANDING ON THE DECK OF THE BRIDGE I hear the wind as it flows around the cables and sets them quivering. I can sense slow movements in the deck and handrail as heavy vehicles pass. When I touch one of the cables I can feel it heating up in the sunlight and imagine it lengthening slightly, bending the towers and lowering the deck.

The bridge seems alive – an elastic web of steel and concrete in constant motion. Like a spider's web, it stretches, bends and twists as the wind, sun, rain or traffic encounter it. Like a web, its elasticity and momentum set it trembling, twisting and rippling after it is pushed or pulled out of shape.

In bridges, as in all large structures, these motions must be foreseen, their effects estimated and allowed for in their design and construction. Bridges, unlike buildings, are not enclosed structures. Their nakedness reveals to all the solution worked out by their designers to ensure stability in all imaginable circumstances during construction and use.

The challenge presented to engineers at the Roads and Traffic Authority (RTA) in the mid-1980s was to find an economical replacement for the old Glebe Island Bridge which, after operating since 1903, was nearing the end of its useful life. The new bridge would need to carry six traffic lanes and a pedestrian lane at a level high enough to allow clearance for ships entering or leaving Blackwattle Bay and Rozelle Bay.

The initial plans were for a concrete box girder bridge with a main span of 200 metres supported by piers in the water at the entrance to Johnsons Bay. Piles would be driven into the seabed around each of these piers to protect them from damage by ships.

However when geotechnical engineers analysed the likely effects of a 10 000 tonne vessel hitting the piles at 8 knots, they concluded that the movement of the seabed could rupture nearby electricity and telephone cables – an event which would cause chaos in the Central Business District and inner suburbs.

After considering several options (including relocating the cables), the RTA engineers concluded that the most cost-effective solution was to increase the length of the bridge's main span so that the piers at each end would be out of the water. This would require a main span of over 300 metres, well beyond the maximum span considered feasible for a box girder bridge. A concrete arch bridge like the Gladesville Bridge would have required strong rock footings to withstand the large outward thrust at each end of the arch, but these were not available at the Glebe Island site.

Instead, the RTA engineers proposed a cable-stayed design for the main span: cables from two high towers would balance the weight of the concrete deck in the main span with the weight of the decks in the adjoining spans which would be linked to the roads on either side. In recent years engineers in many other countries had chosen cable-stayed designs for long spans but this would represent the latest developments in this construction technology in Australia.

After visits overseas to inspect some recently completed cable-stayed bridges, RTA engineers developed a new design for the Glebe Island site. Unlike the previous box girder design, the cable-stayed bridge would require tall towers and fan-like sets of cables on a scale which would dominate the local skyline. A supplementary Environmental Impact Statement assessed the appearance of the new design and subsequently some changes were made to enhance its visual appeal.

The final specifications drawn up by the RTA engineers included a main span of 345 metres with two adjoining spans each of 140 metres. These spans would be supported by 128 stay cables fixed to two 120 metre high delta-shaped concrete towers. One approach span at the Pyrmont end and two at the White Bay end would link the bridge to new roadways. These three approach spans would incorporate conventional concrete box girders.

The final designs specified a concrete deck 32 metres wide and 25 centimetres deep with concrete beams under each side and a cross girder every five metres. The towers and the deck with its beams and girders would be made of reinforced concrete, cast in place and post-stressed to withstand the bending forces caused by their own weight and the weight of vehicles on the deck.

To develop and check these designs the RTA engineers used space frame analysis computer programs to calculate forces in the bridge structure in different conditions. In these programs the calculations are made on the assumption that the steel and concrete components all stretch and bend elastically. Allowance was made for any one of the cables to be removed or stretched beyond its limit of elasticity without putting an unsafe load on the others.

Other kinds of design testing were done to check the bridge's dynamic behaviour in high winds. Like a spider's web which quivers and ripples in the wind, bridge decks can vibrate and twist in very high winds. If these two dynamic behaviours occur together, potentially dangerous "flutter" can occur. Wind tunnel testing of a model of the bridge and a three-dimensional computer simulation of the bridge's dynamics indicated that vibration and twisting would occur at very different frequencies and so would not cause the flutter which is known to have weakened and destroyed other bridges.

After finalising the design and specifications, the RTA invited experienced companies to tender for the various tasks in the construction project. The first major task was to construct the foundations for the two towers. Each tower sits on a massive concrete pile cap joined to the tops of 56 underground piles. These piles – cylinders of reinforced concrete 1.5 metres in diameter – distribute the weight of the bridge onto the sandstone bedrock below.

Leighton Contractors was awarded the contract to construct the tower piles and pile caps, and Morson Engineering of Wyong the contract to manufacture the steel casings for the 112 piles.

In 1989 the steel casings were delivered. Later that year the first were lifted into an upright position and, over several hours, driven through the soil and clay by repeated hits from a pneumatic hammer. After reaching the bedrock (in some cases 35 metres below), the soil and clay inside were removed, steel reinforcing put in place and concrete poured in and compacted. The protruding part of each casing was then cut off.

The pile caps which support the bases of the towers each required 400 tonnes of steel reinforcing and a volume of concrete equal to the water in three Olympic-size swimming pools. These huge concrete pours were completed late in 1990.

But even before this happened, the New South Wales Government, responding to pressure to reduce expenditure in the recession years of the early 1990s, had decided to delay the remaining work on the project. Work did not resume until after the Federal Government announced early in 1992 as part of its One Nation statement that it would pay some of the construction costs as part of its program to provide more jobs.

The main contract in the bridge project was for construction of the towers and deck. This was won by Baulderstone Hornibrook, with Austress Freyssinet winning the subcontract for supplying and installing the stay cables.

The design for the completed bridge was complex – having to take into account the forces on thousands of interconnected components made of many different materials in all imaginable circumstances. But even more complex were the design checks needed to ensure that the structure would be stable at all stages during its construction. Every single addition to the bridge structure during construction was in effect one more structure which had to be analysed and checked to ensure stability and safety.

So important was this consideration that the construction contract included in its specifications a requirement that the contractor engage a specialist engineering company to check that the bridge structure would be stable at all stages during its construction. Baulderstone Hornibrook engaged DRC Consultants of New York, a company which had planned and checked construction sequences on several cable-stayed bridges in other countries, using a computer-based model to predict how forces on different components of a bridge change at each step in its construction.

The next stage of the project – construction of the western tower and piers for the approach spans – began in mid-1992. Temporary formwork was erected, steel reinforcing fabricated and put in position, then concrete poured and compacted. This cycle of work was completed many times to make the tower base and then the lower outwards-sloping tower legs.

Falsework using Favco support towers and steel girders was erected to support the formwork for the horizontal tie beam which connects the lower legs of the tower just below deck level. At first glance this beam appears to support the weight of the deck,

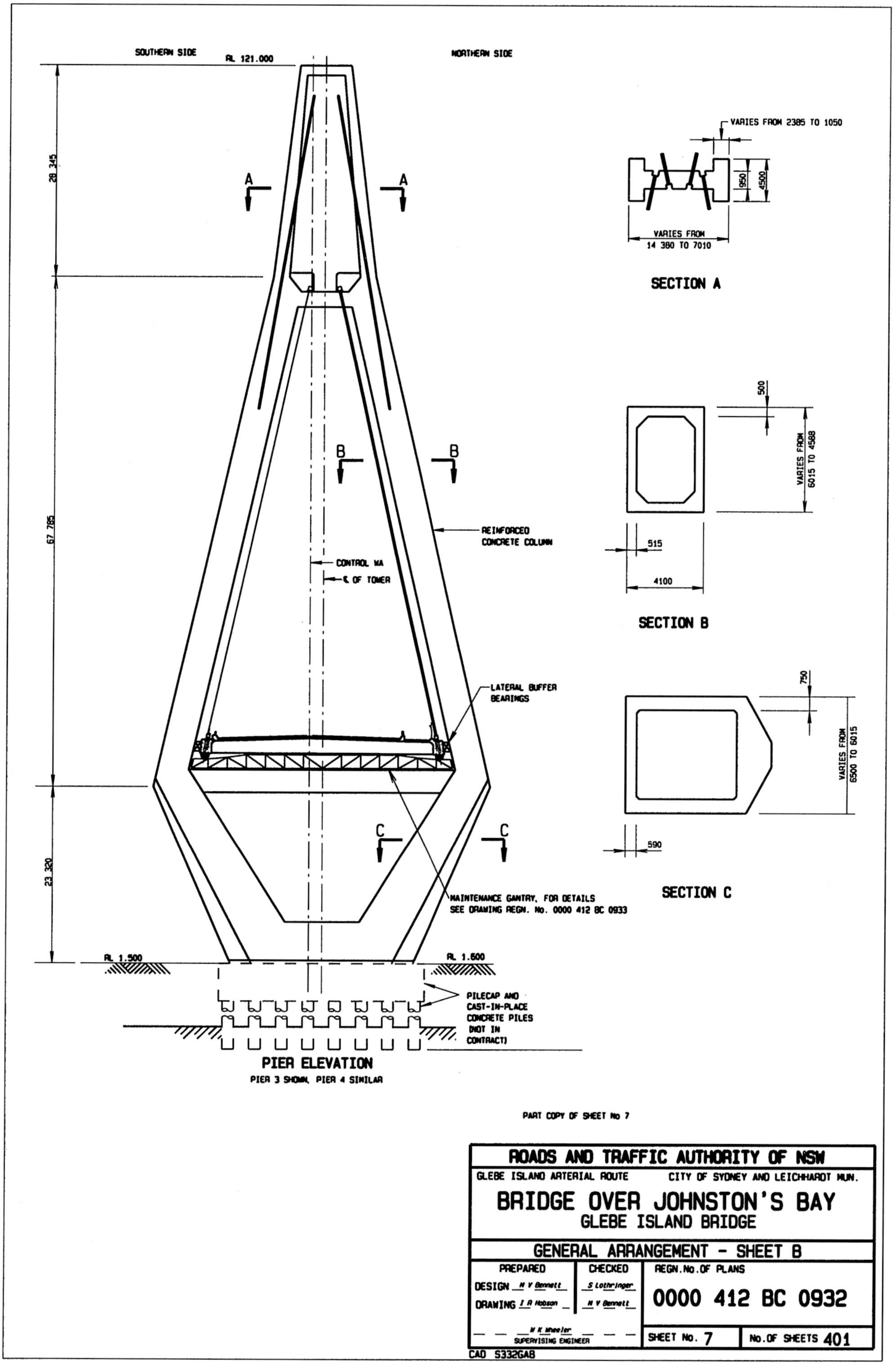

SOUTHERN SIDE
RL 121.000
NORTHERN SIDE
28 345
67 785
23 320
A
B
C
REINFORCED CONCRETE COLUMN
CONTROL WA
℄ OF TOWER
LATERAL BUFFER BEARINGS
MAINTENANCE GANTRY, FOR DETAILS SEE DRAWING REGN. No. 0000 412 BC 0933
RL 1.500
RL 1.600
PILECAP AND CAST-IN-PLACE CONCRETE PILES (NOT IN CONTRACT)
PIER ELEVATION
PIER 3 SHOWN, PIER 4 SIMILAR
VARIES FROM 2385 TO 1050
950
4500
VARIES FROM 14 380 TO 7010
SECTION A
500
VARIES FROM 6015 TO 4588
515
4100
SECTION B
750
VARIES FROM 6500 TO 6015
590
SECTION C
PART COPY OF SHEET No 7
ROADS AND TRAFFIC AUTHORITY OF NSW
GLEBE ISLAND ARTERIAL ROUTE
CITY OF SYDNEY AND LEICHHARDT MUN.
BRIDGE OVER JOHNSTON'S BAY
GLEBE ISLAND BRIDGE
GENERAL ARRANGEMENT - SHEET B
PREPARED
CHECKED
REGN. No. OF PLANS
DESIGN H V Bennett
S Lothringer
DRAWING I R Hobson
H V Bennett
0000 412 BC 0932
W K Wheeler
SUPERVISING ENGINEER
SHEET No. 7
No. OF SHEETS 401
CAD S332GAB

but a second look reveals a gap between the deck and the beam – the cables completely support the weight of the deck and the traffic on it.

The upper legs of the delta-shaped towers were constructed in 4-metre sections. Each section required a cycle of activities which began with jacking up a steel-frame "jump-form" bolted to the tower surfaces below, constructing formwork, welding and positioning steel reinforcing inside it, pouring concrete and compacting it to ensure an even mix and uniform strength after setting.

One of the early tasks for DRC Consultants was to design the temporary bracing needed to join and stiffen the separate upper legs during construction – before they were permanently joined at the tower head. Although the tower legs look monolithic, they are actually hollow with half-metre thick walls, allowing inside access to the tower heads for inspection and maintenance of the cables and lights.

While the tower head was being constructed, the first three segments of the deck (called the pier table) were cast in place, supported by Kingshore scaffolding erected around the base of the tower. After the concrete in each segment had set, it was post-tensioned (to squeeze the concrete) by using jacks to stretch the steel tendons which extend through the concrete deck segments, cross girders and side beams. The tendons are held in their stretched state by injecting grouting which binds the tendons to the concrete around them when it sets.

After the tower head had been constructed in mid-1993, the first four pairs of cables could be installed. Each deck segment is supported at its leading corners (on the edge furthest from the tower) by a pair of cables anchored at one end in a recess in the tower head and at the other end in a pod under the deck side beams.

Like the threads in a spider's web, the stay cables are made of many individual strands. Each strand is a bundle of seven galvanised steel wires, packed in wax and covered in a plastic sheath. The number of strands in each cable varies from 25 for deck segments near the towers to 74 for the segments at mid-span, bundled together and enclosed in a black polyethylene pipe.

Despite their real differences in size, the cables all appear to be the same diameter when viewed from a distance – an illusion of constancy which enhances our appreciation of the form of the bridge.

Installing the first cable involved a sequence of several steps, which were to be repeated for each of the 128 cables on the bridge. The first strand in the cable was cut to its required length, drawn through the pipe up to the tower head, pulled by a hydraulic jack in the tower head to the estimated tension force needed, and its ends wedged tight. The other strands in the cable were pulled through and secured in turn, then individually adjusted to bring them all to the same tension.

While work on the western tower proceeded, the two approach spans at the western end were completed. Here the deck incorporates concrete box girders which were cast in place on formwork held up by the permanent piers and by temporary trusses. Reinforced Earth concrete panels were used to make retaining walls at the abutments where the elevated deck meets the new roadway leading to White Bay.

With the western tower, pier table and approach spans complete, the next and most challenging construction stage could begin. This was the construction of the western half of the cable-stayed deck, from the western tower east over water towards the bridge centre, and west over land to join the deck on the completed approach spans.

The procedure allowed every 10-metre deck segment cast on one side of the tower to be balanced by an equal segment cast on the other side. On the water side of the tower the formwork for casting new segments was provided by a moveable 'formtraveller' temporarily attached to the leading pair of cables and to the previous segment. On the land side the formwork was supported by a 'rolling frame' which was designed to be easily moved to the next position for casting new segments.

The steel-frame formtraveller temporarily supported by cables at its leading edge was a recent innovation in bridge construction which had been introduced by DRC Consultants at another cable-stayed bridge in 1992. For the Glebe Island Bridge, refinements made by DRC Consultants to the design of the segments and of the formtraveller allowed much larger segments of about 460 tonnes each to be cast, reducing the number of segments needed, thus shortening the total construction time.

The rolling frame, an interconnected grid of Kingshore towers which could be quickly lowered and moved along on rails to support the formwork for the next two land-side segments, was an innovation developed jointly by DRC Consultants and Baulderstone Hornibrook. By reducing the time to prepare the formwork for each segment, it also significantly reduced the total time taken to construct the deck. In addition, the rolling frame anchored the incomplete land-side deck to the ground and eliminated the need for buffeting cables to prevent it swaying in high winds.

The first tasks in this new stage of construction were to jack up and attach the formtraveller (to the water side) and the rolling framework (to the land side) of the pier table. The deck construction could then proceed in cycles in which the main activities were to:

- advance the formtraveller to cast the next segment;
- place counterweights on the land-side leading segment, install the reinforcing and cast the new land-side segment;
- install and tension new cables attached to the formtraveller;
- post-tension the tendons in the new land-side segment;
- remove the counterweights on the previous land-side segment;

- advance the rolling frame to the next casting position;
- install and tension the cables for the new land-side segment;
- install the reinforcing and cast a new segment on the formtraveller;
- post-tension the tendons in the new water-side segment; and
- transfer the leading water-side cable anchorages from the formtraveller to the pods under the deck side beams.

At several steps in this cycle additional forces such as the weight of new concrete would stretch the cables and bend the tower and deck slightly. After these steps DRC would use its computer-based model of the partial bridge structure to calculate the adjustments to cable tensions needed to bring the structure back to its preferred shape.

Early in this stage of construction, temporary buffeting cables were installed between the base of the tower and the incomplete water-side deck to control swaying in high winds. Regular surveys were done to check and correct any significant deviations from the theoretically predicted shape of the partial bridge structure.

By mid-1994 the western half of the cabling and deck were complete and joined at the White Bay end to the deck over the two approach spans. Meanwhile the eastern tower and pier table and the eastern approach span had been completed following the same construction sequence used on the western side. The 260 tonne formtraveller had been jacked down onto a barge, floated across to the eastern tower, jacked up and attached to the water side of the eastern pier table.

Construction of the eastern half of the deck and cabling followed the same cycle of activities used for the western half except that, because of the roads in use below, the rolling frame could not be used to support the formwork for all of the land-side segments. Instead the formwork was supported on a temporary framework of steel towers and trusses which provided sufficient clearance for traffic using the old bridge.

By June 1995 the eastern half of the cabling and deck were complete and joined at its east end to the deck over the Pyrmont approach span, leaving a small gap between the two halves of the deck. On 24 July the separate halves were clamped with temporary beams to hold them together while the closing segment was cast and post-stressing completed. To minimise the movements caused by the cables expanding and contracting, the closing segment was cast at night when the air and cable temperatures are most stable.

While the road surfaces were being laid and the railings and lights installed, the formtraveller and other temporary supports were removed. The complete bridge structure was checked and surveyed, and the cable lengths adjusted at their pod anchorages to pull the deck into its correct curved profile. The new bridge was opened to traffic on 4 December 1995.

Pedestrians walking over the new bridge can catch glimpses of the Gladesville Bridge and Sydney Harbour Bridge, and see the old Glebe Island Bridge below. These four bridges, built at approximately 30-year intervals during this century, all represent state-of-the-art bridge design and construction for their respective eras.

Each one challenged and extended its designers and builders, and each prompted public concerns about its necessity, feasibility and appearance. The new Glebe Island Bridge, like the other three bridges before it, will be valued for its innovations in design and construction, its convenience and its presence in the urban landscape around it.

Robert Renew is curator of engineering design at the Powerhouse Museum, Sydney.

Crane swings forklift truck off deck

Western tower with mobile crane wheel

Steelfixer Richard Harris tying reinforcing bars in preparation for concrete placement

Clockwise from top left: Tony Sutherland driving the Alimak hoist on western tower; Astrubel Eusebio, leading hand carpenter; Trevor Weller, foreman; Mario Da Conceicao, carpenter

Kofeinu Mataresi, concrete worker

Opposite page: New and old Glebe Island Bridges with narrowing gap between concrete decks

THINK BIG.

East and west decks close to joining as seen from Glebe Point

 Placing concrete segment on eastern deck. Total bridge span has 53 concrete segments

Concrete workers in cold winter conditions

 Concrete workers with Harbour Bridge in background

Mick Scully, concrete pump operator

 Henry Figueira screeding concrete on southern side of deck

Henry Figueira's face shows exhaustion during concreting work

Dawn light illuminating the bridge as deck gap narrows. Glebe Island shipping terminal in background ▶

Preparing for final two metre concrete placement on 24 July 1995 with old Glebe Island Bridge below
Opposite page: Storm clouds sweep over western face of city past bridge stay cables and completed towers

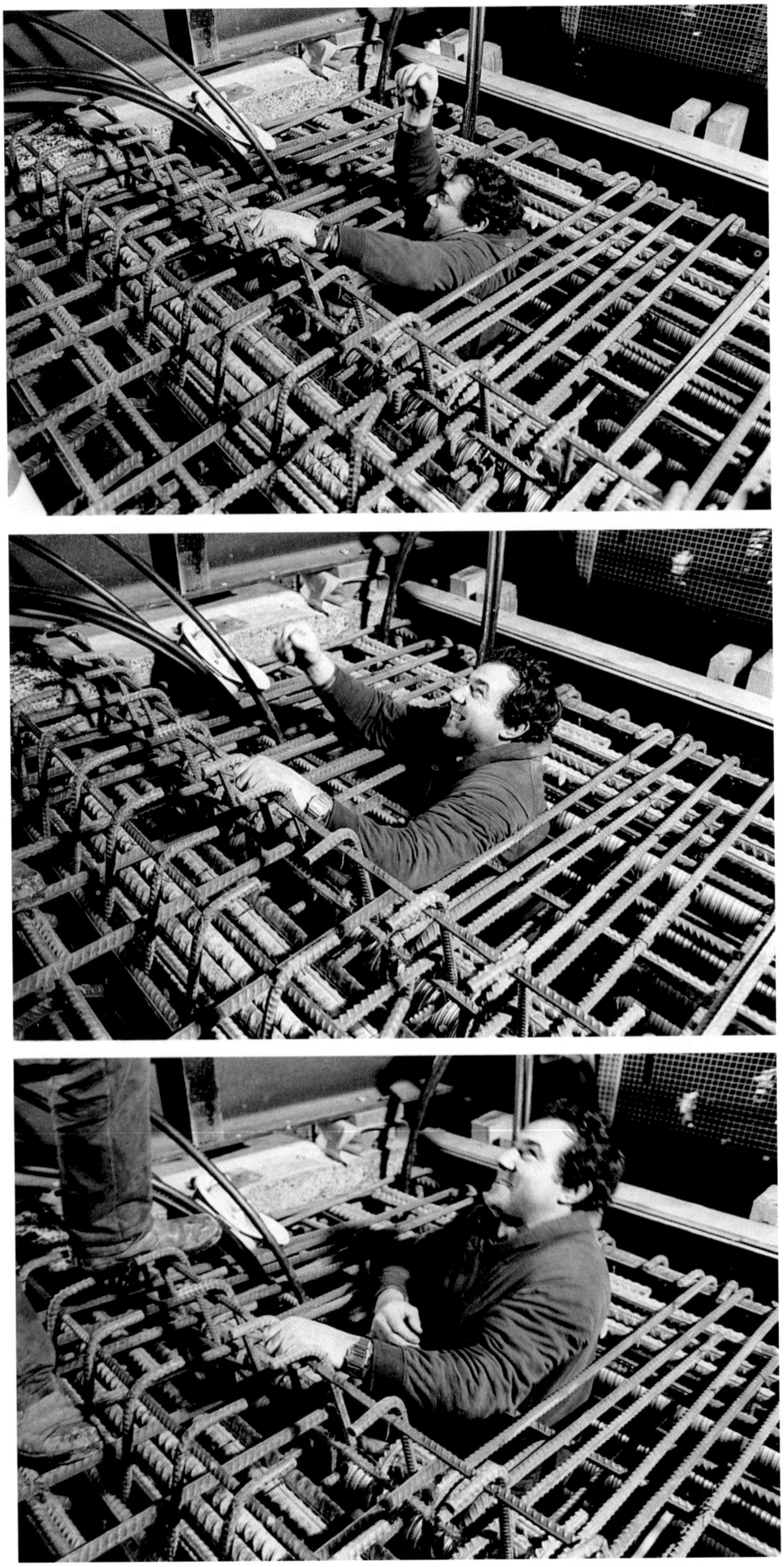

Carlos Avelar emerges from reinforcing steel assembly after cleanout before final concrete placement

Opposite page: Night concrete placement of final two-metre gap between decks

Above & opposite page: Baulderstone Hornibrook workers relaxing in site office after completion of final two-metre concrete segment on bridge deck

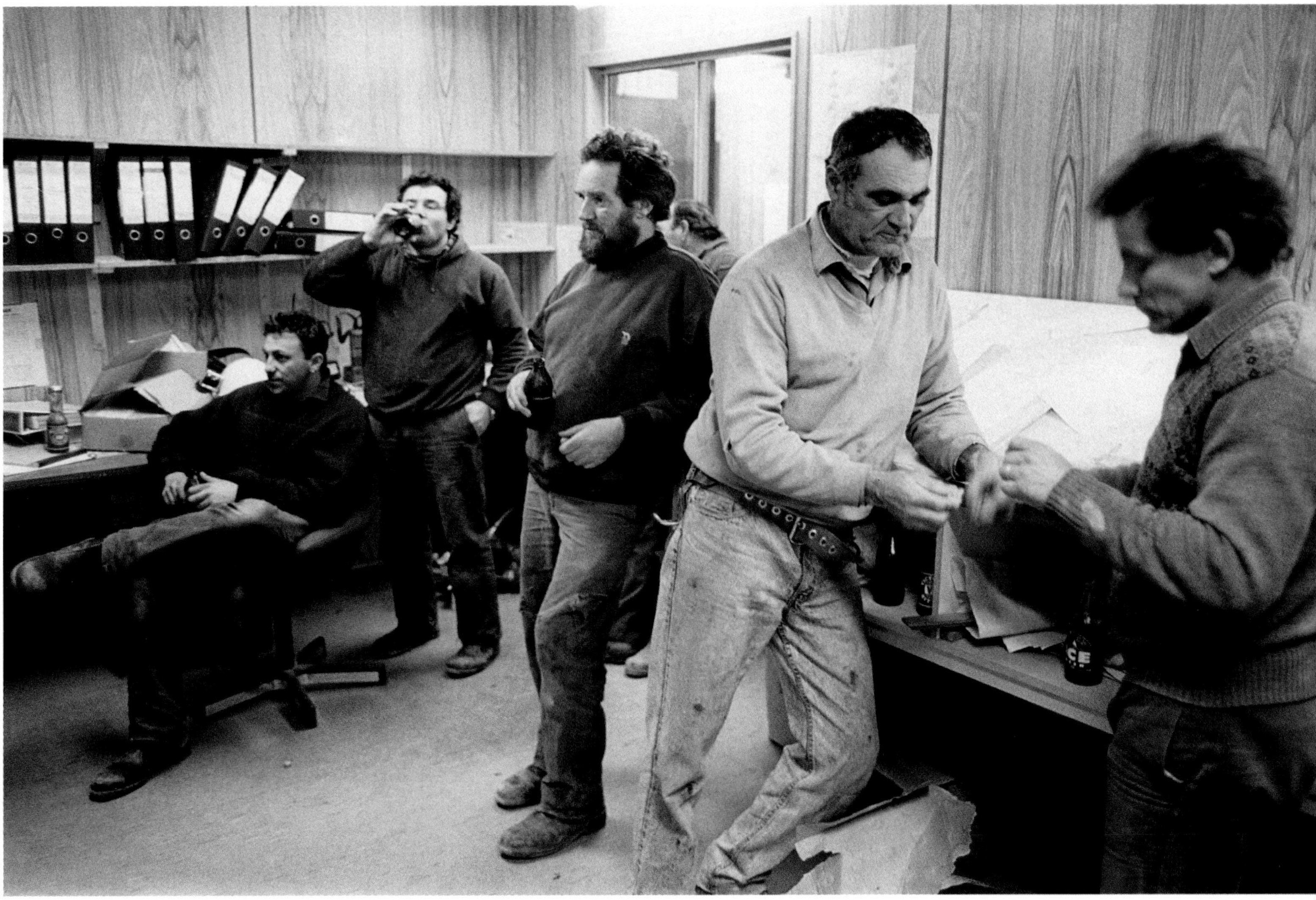

 The bridge at dawn from the head of Rozelle Bay

AIDC

Workers returning in Alimak hoist to ground level. *Top:* Renzo Calligaro (centre) and Barney Kreft (right). *Bottom left to right:* John Griffith, Angelo Navrazoglou and Nigel Barker
Opposite page: The gap is closed on the segmented deck

 A protected fascia panel is lifted to position before fixing on western tower head

Barney Kreft reaches out to guide a fascia panel rising towards tower head

An aluminium fascia panel is guided inwards as the protective cover is removed before bolting to framework
Opposite page: Barney Kreft guides a fascia panel into place

Tower head worker's gloves

◀ Dusk light reflects from city buildings behind the completed main span

Spirit level and reinforcing rods

General view onto western approach span

Opposite page: Formtraveller is lowered onto barges after completion of western deck segments

Sunrise on the bridge looking south-west

ING Mercantile Mutual Bank
Are you flying

Colour-coded outer stay cable sheaths. Colours denote differing diameters

Welding outer stay cable sheaths into continuous lengths

The bridge deck near joining with the old Glebe Island Bridge and shipping terminal below ▶

BRIDGING TOWARDS
SYDNEY 2000
TEXAS

Construction graphic and steel detail

Opposite page: Late sun illuminating western tower after passing storm

1348·67

SKATE
BEAM

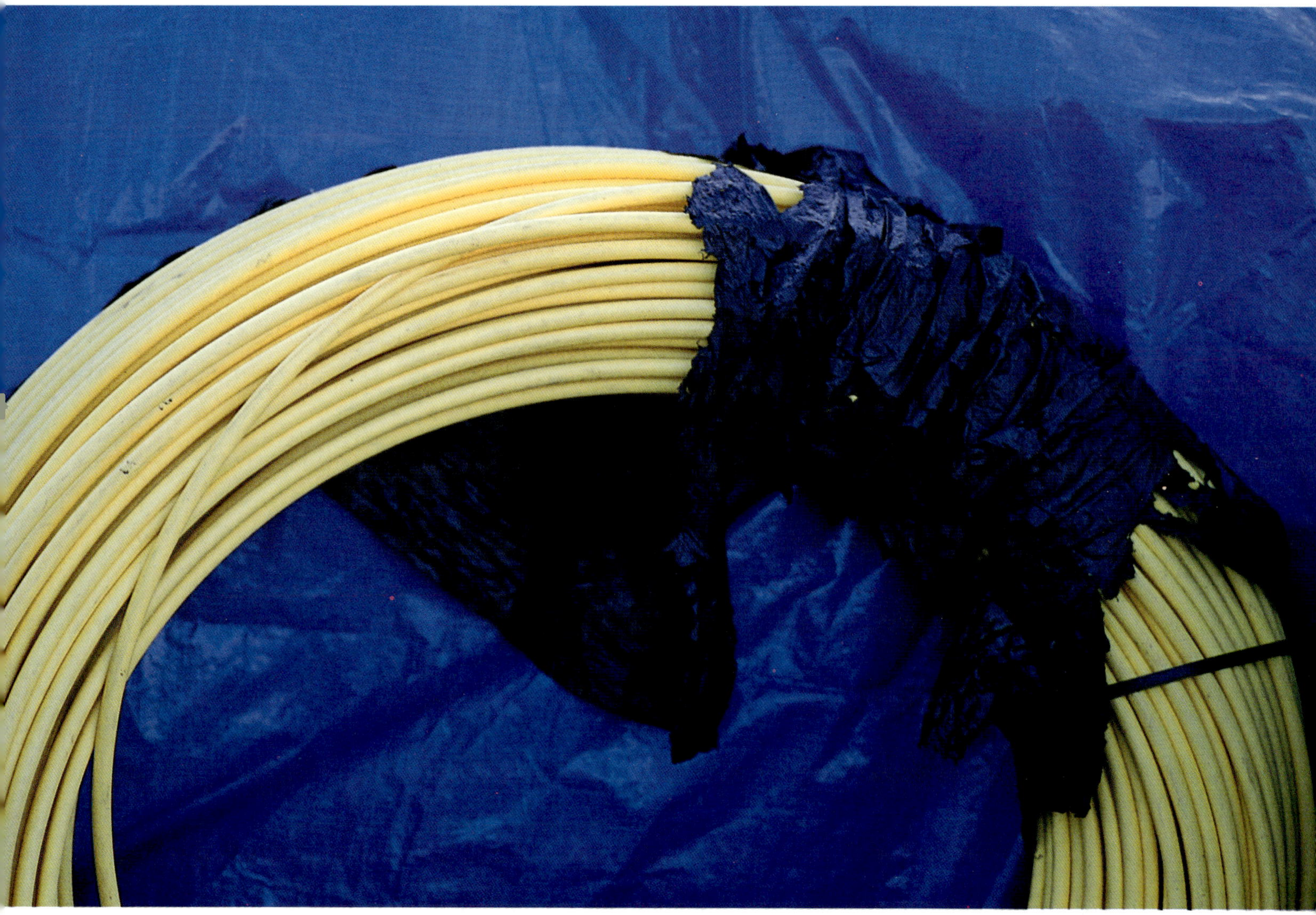

Tubing used in concrete placement on deck slab

The full moon rises over the western face of the city with bridge in late stage of construction ▶

National Mutual

PACIFIC POWER
J.&J. ROBERTSON & SONS Pty. Ltd. • General Carriers • Container Haulage 818 1011
HYMIX

 Tool box on deck

From left: Galvanised steel maintenance gantry beams support brackets; aluminium pedestrian barrier rail sections await placement and fixing

 Tensioned stay cables

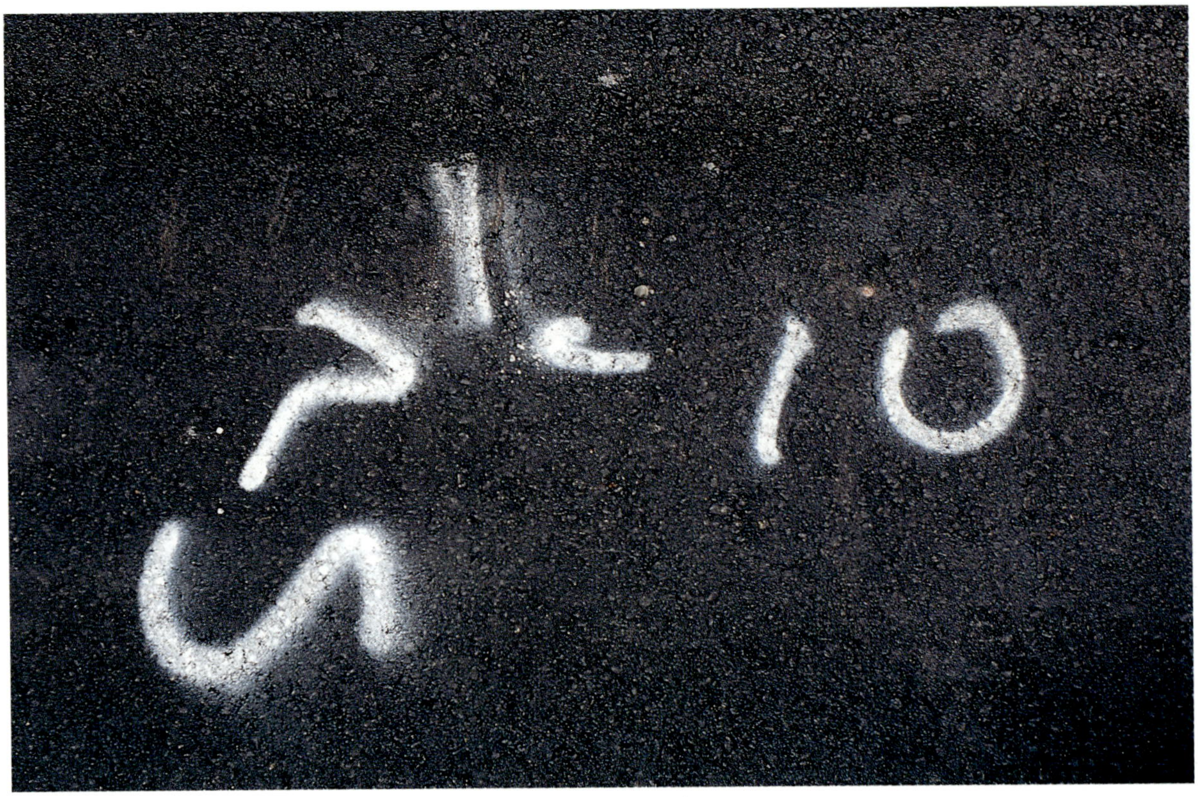

 Graphic on asphalt

Greg Cooney, Austress Freyssinet leading hand, using mono-strand jack for final tensioning of cable strands in anchorage at top of western tower

DO NOT

Bernie Newell rolling aggregate into bitumen seal
Opposite page: Steam cloud rises as hot bitumen is sprayed onto deck

J.&J. ROBERTSON & SONS Pty. Ltd. General Carriers - Container Haulage 818 1011

Guests on approach deck at gala dinner the evening before the official opening
Opposite page: Lightning strike on Centrepoint Tower seen from Rozelle

 Fireworks and floodlit towers at celebration dinner

From top: New South Wales Premier Bob Carr making opening speech before cutting ribbon on 3 December 1995. NSW Minister for Transport Michael Knight at left of dais; Premier Carr and Minister Knight cross the bridge after opening ceremony

Above & opposite page: Part of crowd of 65,000 who walked across the completed bridge on opening day. The start was carefully staggered to avoid a crush

People getting a low level view of western tower and cables from roadway
Opposite page: A dancing couple under western tower

Looking east across the newly-completed bridge to the city of Sydney with the Opera House and Harbour Bridge (top left)

Traffic lights trace the curve of the roadway at dusk ▶

SEE THE U.S. ON U.S.

The eastern tower head against the western face of the city